An Introduction to Steam Turbine Design

2nd Edition

J. Paul Guyer, P.E., R.A.

Editor

The Clubhouse Press
El Macero, California

CONTENTS

1. TYPICAL PLANTS AND CYCLES
2. COGENERATION IN STEAM POWER PLANTS
3. TURBINE TYPES
4. TURBINE GENERATOR SIZES
5. TURBINE THROTTLE PRESSURE AND TEMPERATURE
6. TURBINE EXHAUST PRESSURE
7. LUBRICATING OIL SYSTEMS
8. GENERATOR TYPES
9. GENERATOR COOLING
10. TURBINE GENERATOR CONTROL
11. TURNING GEAR
12. TURBINE GENERATOR FOUNDATIONS
13. AUXILIARY EQUIPMENT
14. INSTALLATION
15. CLEANUP, STARTUP, AND TESTING
16. OPERATION

(This publication is adapted from the *Unified Facilities Criteria* of the United States government which are in the public domain, have been authorized for unlimited distribution, and are not copyrighted.)

(Figures, tables and formulas in this publication may at times be a little difficult to read, but they are the best available. **DO NOT PURCHASE THIS PUBLICATION IF THIS LIMITATION IS UNACCEPTABLE TO YOU.**)

1. TYPICAL PLANTS AND CYCLES

1.1 DEFINITION. The cycle of a steam power plant is the group of interconnected major equipment components selected for optimum thermodynamic characteristics, including pressures, temperatures, and capacities, and integrated into a practical arrangement to serve the electrical (and sometimes by-product steam) requirements of a particular project. Selection of the optimum cycle depends upon plant size, cost of money, fuel costs, non-fuel operating costs, and maintenance costs.

1.2 STEAM TURBINE PRIME MOVERS

1.2.1 SMALLER TURBINES. Turbines under 1,000 kW may be single stage units because of lower first cost and simplicity. Single stage turbines, either back pressure or condensing, are not equipped with extraction openings.

1.2.2 LARGER TURBINES. Turbines for 5,000 kW to 30,000 kW shall be multi-stage, multi-valve units, either back pressure or condensing types.

1.2.2.1 BACK PRESSURE TURBINES. Back pressure turbine units usually exhaust at pressures between 5 psig (34 kPa gage) and 300 psig (2068 kPa gage) with one or two controlled or uncontrolled extractions. However, there is a significant price difference between controlled and uncontrolled extraction turbines, the former being more expensive. Controlled extraction is normally applied where the bleed steam is exported to process or district heat users.

1.2.2.2 CONDENSING TURBINES. Condensing units exhaust at pressures between 1 inch of mercury absolute (Hga) and 5 inches Hga, with up to two controlled, or up to five uncontrolled extractions.

1.3 SELECTION OF CYCLE CONDITIONS. The function or purpose for which the plant is intended determines the conditions, types, and sizes of steam generators and turbine drives and extraction pressures.

1.3.1 SIMPLE CONDENSING CYCLES. Straight condensing cycles or condensing units with uncontrolled extractions are applicable to plants or situations where security or isolation from public utility power supply is more important than lowest power cost. Because of their higher heat rates and operating costs per unit output, it is not likely that simple condensing cycles will be economically justified for some power plant applications as compared with that associated with public utility purchased power costs. A schematic diagram of an uncontrolled extraction-cycle is shown in Figure 1.

1.3.2 CONTROLLED EXTRACTION-CONDENSING CYCLES AND BACK PRESSURE CYCLES. Back pressure and controlled extraction-condensing cycles are attractive and applicable to a cogeneration plant, which is defined as a power plant simultaneously supplying either electric power or mechanical energy and heat energy. A schematic diagram of a controlled extraction-condensing cycle is shown in Figure 2. A schematic diagram of a back pressure cycle is shown in Figure 3.

1.3.3 TOPPING CYCLE. A schematic diagram of a topping cycle is shown in Figure 4. The topping cycle consists of a high pressure steam boiler and turbine generator with the high pressure turbine exhausting steam to one or more low pressure steam turbine generators. High pressure topping turbines are usually installed as an addition to an existing lower pressure steam electric plant.

1.4 GENERAL ECONOMIC RULES. Maximum overall efficiency and economy of the steam turbine power cycle are the objectives of a satisfactory design. Higher efficiency and a lower heat rate require more complex cycles which are accompanied with higher initial investment costs and higher operational and maintenance costs but lower fuel costs. General rules to consider to improve the plant efficiency are listed hereinafter.

a) Higher steam pressures and temperatures increase the turbine efficiencies, but temperatures above 750 degrees F (399 degrees C) usually require more expensive alloy piping in the high pressure steam system.

b) Lower condensing pressures increase turbine efficiency. However, there is a limit where lowering condensing (back) pressure will no longer be economical, because the costs of lowering the exhaust pressure is more than the savings from the more efficient turbine operation.

c) The use of stage or regenerative feedwater cycles improves heat rates, with greater improvement corresponding to larger numbers of such heaters. In a regenerative cycle, there is also a thermodynamic crossover point where lowering of an extraction pressure causes less steam to flow through the extraction piping to the feed water heaters, reducing the feedwater temperature. There is also a limit to the number of stages of extraction/feedwater heating, which may be economically added to the cycle. This occurs when additional cycle efficiency no longer justifies the increased capital cost.

d) Larger turbine generator units are generally more efficient than smaller units.

e) Multi-stage and multi-valve turbines are more economical than single stage or single valve machines.

f) Steam generators of more elaborate design and with heat saving accessory equipment are more efficient.

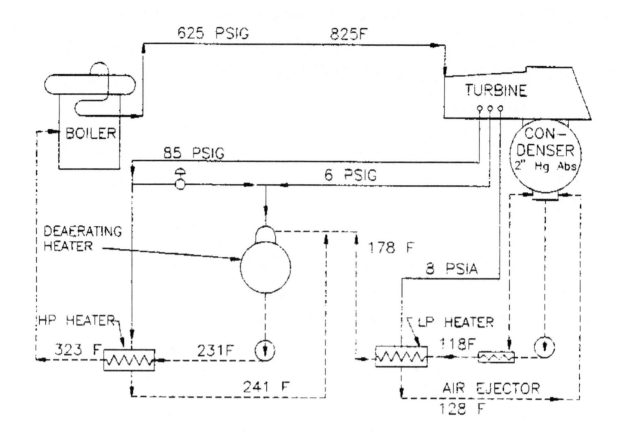

Figure 1

Typical uncontrolled extraction – condensing cycle

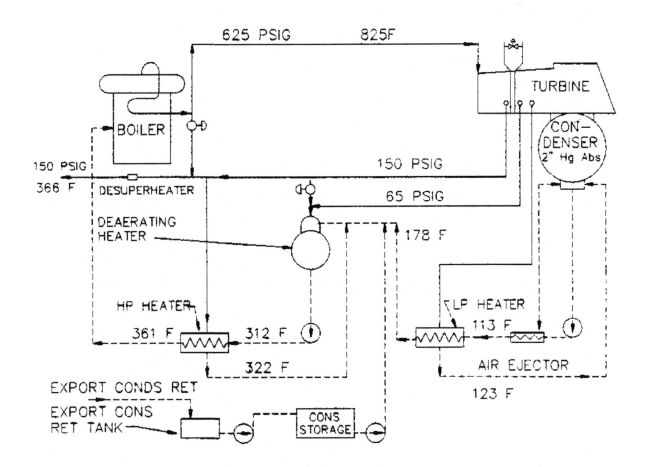

Figure 2

Typical controlled extraction – condensing cycle

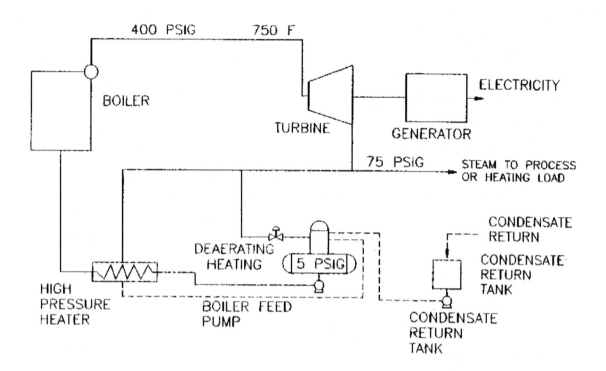

Figure 3
Typical back pressure cycle

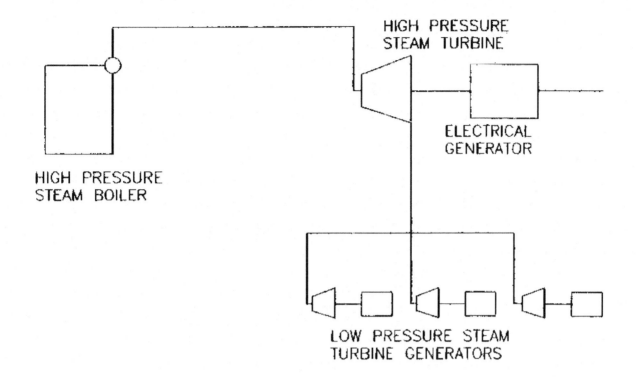

Figure 4
Typical topping cycle

1.5 SELECTION OF CYCLE STEAM CONDITIONS

1.5.1 BALANCED COSTS AND ECONOMY. For a new or isolated plant, the choice of initial steam conditions should be a balance between enhanced operating economy at higher pressures and temperatures, and generally lower first costs and less difficult operation at lower pressures and temperatures. Realistic projections of future fuel costs may tend to justify higher pressures and temperatures, but such factors as lower availability, higher maintenance costs, more difficult operation, and more elaborate water treatment shall also be considered.

1.5.2 EXTENSION OF EXISTING PLANT. Where a new steam power plant is to be installed near an existing steam power or steam generation plant, careful consideration shall be given to extending or paralleling the existing initial steam generating conditions. If existing steam generators are simply not usable in the new plant cycle, it may be appropriate to retire them or to retain them for emergency or standby service only. If boilers are retained for standby service only, steps shall be taken in the project design for protection against internal corrosion.

1.5.3 SPECIAL CONSIDERATIONS. Where the special circumstances of the establishment to be served are significant factors in power cycle selection, the following considerations may apply:

1.5.3.1 ELECTRICAL ISOLATION. Where the proposed plant is not to be interconnected with any local electric utility service, the selection of a simpler, lower pressure plant may be indicated for easier operation and better reliability.

1.5.3.2 GEOGRAPHIC ISOLATION. Plants to be installed at great distances from sources of spare parts, maintenance services, and operating supplies may require special consideration of simplified cycles, redundant capacity and equipment, and highest practical reliability. Special maintenance tools and facilities may be required, the cost of which would be affected by the basic cycle design.

1.5.3.3 WEATHER CONDITIONS. Plants to be installed under extreme weather conditions require special consideration of weather protection, reliability, and redundancy. Heat rejection requires special design consideration in either very hot or very cold weather conditions. For arctic weather conditions, circulating hot water for the heat distribution medium has many advantages over steam, and the use of an antifreeze solution in lieu of pure water as a distribution medium should receive consideration.

1.6 STEAM POWER PLANT ARRANGEMENT

1.6.1 GENERAL. Small units utilize the transverse arrangement in the turbine generator bay, while the larger utility units are very long and require end-to-end arrangement of the turbine generators.

1.6.2 TYPICAL SMALL PLANTS. Figures 5 and 6 show typical transverse small plant arrangements. Small units less than 5,000 kW may have the condensers at the same level as the turbine generator for economy, as shown in Figure 5. Figure 7 indicates the critical turbine room bay clearances.

1.7 HEAT RATES. The final measure of turbine cycle efficiency is represented by the turbine heat rate. It is determined from a heat balance of the cycle, which accounts for all flow rates, pressures, temperatures, and enthalpies of steam, condensate, or feedwater at all points of change in these thermodynamic properties. Heat rate is an excellent measure of the fuel economy of power generation.

1.7.1 HEAT RATE UNITS AND DEFINITIONS. The economy or efficiency of a steam power plant cycle is expressed in terms of heat rate, which is total thermal input to the cycle divided by the electrical output of the units. Units are Btu/kWh.

a) Conversion to cycle efficiency, as the ratio of output to input energy, may be made by dividing the heat content of one kWh, equivalent to 3412.14 Btu by the heat rate, as defined. Efficiencies are seldom used to express overall plant or cycle performance, although efficiencies of individual components, such as pumps or steam generators, are commonly used.

b) Power cycle economy for particular plants or stations is sometimes expressed in terms of pounds of steam per kilowatt hour, but such a parameter is not readily comparable to other plants or cycles and omits steam generator efficiency.

c) For mechanical drive turbines, heat rates are sometimes expressed in Btu per hp-hour, excluding losses for the driven machine. One horsepower hour is equivalent to 2544.43 Btu.

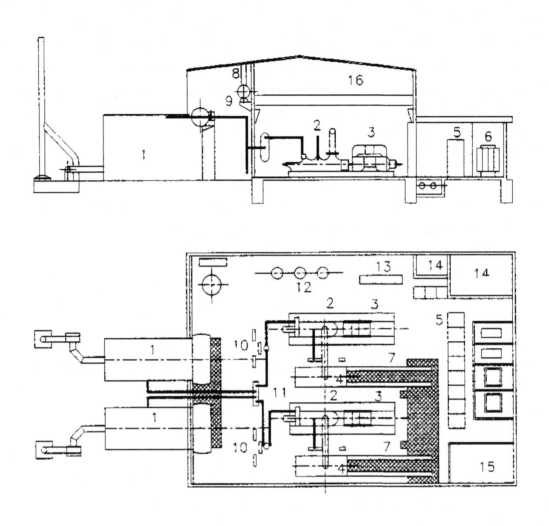

Figure 5

Typical small 2-unit power plant (less than 5 MW), condenser on same level as turbine

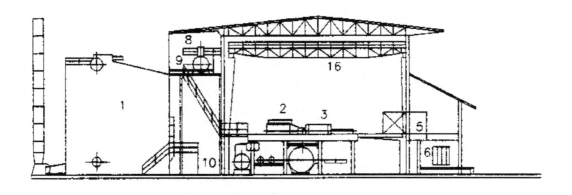
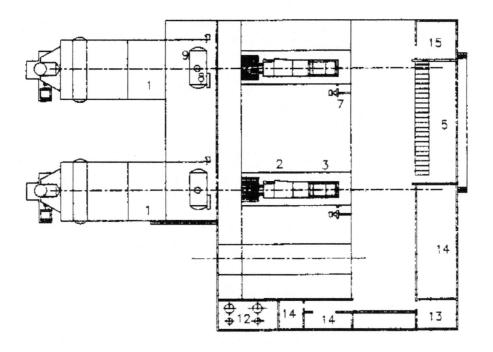

1 – BOILER, OIL FIRED
2 – TURBINE
3 – GENERATOR
4 – CONDENSER
5 – SWITCHGEAR
6 – TRANSFORMERS
7 – COOLING WATER PUMP
8 – DEAERATOR
9 – FEEDWATER TANK
10 – BOILER FEED PUMPS
11 – INDUCED DRAFT FAN
12 – WATER TREATMENT
13 – CHEMICAL LABORATORY
14 – OFFICES
15 – STORE ROOM
16 – TURBINE HOUSE CRANE

Figure 6

Typical 2-unit power plant with condenser below turbine

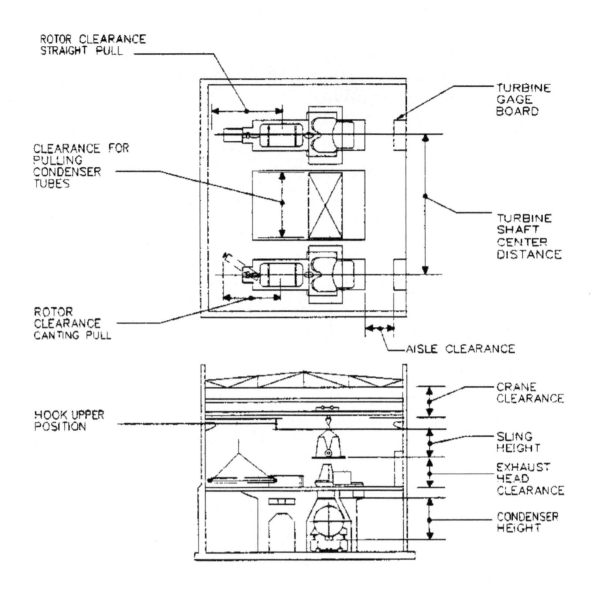

Figure 7

Critical turbine room bay clearances

1.7.2 TURBINE HEAT RATES

1.7.2.1 GROSS TURBINE HEAT RATE. The gross heat rate is determined by dividing the heat added in the boiler between feedwater inlet and steam outlet by the kilowatt output of the generator at the generator terminals. The gross heat rate is expressed in Btu per kWh. For reheat cycles, the heat rate is expressed in Btu per kWh. For reheat cycles, the heat added in the boiler includes the heat added to the steam through the reheater. For typical values of gross heat rate, see Table 4.

Turbine generator rating, kW	Throttle pressure psig	Throttle temperature F deg	Reheat temperature F deg	Pressure In Hg Abs	Cond. Heat rate Btu/kWh
11,500	600	825		1 ½	10,423
30,000	850	900		1 ½	9,462
60,000	1,250	950		1 ½	8,956
75,000	1,450	1,000	1,000	1 ½	8,334
125,000	1,800	1,000	1,000	1 ½	7,904

Table 4

Typical gross turbine heat rates

1.7.2.2 NET TURBINE HEAT RATE. The net heat rate is determined the same as for gross heat rate, except that the boiler feed pump power input is subtracted from the generator power output before dividing into the heat added in the boiler.

1.7.2.3 TURBINE HEAT RATE APPLICATION. The turbine heat rate for a regenerative turbine is defined as the heat consumption of the turbine in terms of "heat energy in steam" supplied by the steam generator, minus the "heat in the feedwater" as warmed by turbine extraction, divided by the electrical output at the generator terminals. This definition includes mechanical and electrical losses of the generator and turbine auxiliary systems, but excludes boiler inefficiencies and pumping losses and loads. The turbine heat rate is useful for performing engineering and economic comparisons of various turbine designs.

1.7.3 PLANT HEAT RATES. Plant heat rates include inefficiencies and losses external to the turbine generator, principally the inefficiencies of the steam generator and piping systems; cycle auxiliary losses inherent in power required for pumps and fans; and related energy uses such as for soot blowing, air compression, and similar services.

1.7.3.1 GROSS PLANT HEAT RATE. This heat rate (Btu/kWh) is determined by dividing the total heat energy (Btu/hour) in fuel added to the boiler by the kilowatt output of the generator.

1.7.3.2 NET PLANT HEAT RATE. This heat rate is determined by dividing the total fuel energy (Btu/hour) added to the boiler by the difference between power (kilowatts/hour) generated and plant auxiliary electrical power consumed.

1.7.4 CYCLE PERFORMANCE. Both turbine and plant heat rates, as above, are usually based on calculations of cycle performance at specified steady state loads and well defined, optimum operating conditions. Such heat rates are seldom achieved in practice except under controlled or test conditions.

1.7.5 LONG TERM AVERAGES. Plant operating heat rates are actual long term average heat rates and include other such losses and energy uses as non-cycle auxiliaries, plant lighting, air conditioning and heating, general water supply, startup and shutdown losses, fuel deterioration losses, and related items. The gradual and inevitable deterioration of equipment, and failure to operate at optimum conditions, are reflected in plant operating heat rate data.

1.7.6 PLANT ECONOMY CALCULATIONS. Calculations, estimates, and predictions of steam plant performance shall allow for all normal and expected losses and loads and should, therefore, reflect predictions of monthly or annual net operating heat rates and costs. Electric and district heating distribution losses are not usually charged to the power plant but should be recognized and allowed for in capacity and cost analyses. The designer is required to develop and optimize a cycle heat balance during the conceptual or preliminary design phase of the project. The heat balance depicts, on a simplified flow diagram of the cycle, all significant fluid mass flow rates, fluid pressures and temperatures, fluid enthalpies, electric power output, and calculated cycle heat rates based on these factors. A heat balance is usually developed for various increments of plant load such as 25, 50, 75, 100 percent and VWO (valves, wide open). Computer programs have been developed which can quickly optimize a particular cycle heat rate using iterative heat balance calculations. Use of such a program should be considered.

1.8 STEAM RATES

1.8.1 THEORETICAL STEAM RATE. When the turbine throttle pressure and temperature and the turbine exhaust pressure (or condensing pressure) are known, the theoretical steam rate can be calculated based on a constant entropy expansion or can be determined from published tables. See Theoretical Steam Rate Tables, The American Society of Mechanical Engineers, 1969. See Table 5 for typical theoretical steam rates.

P_{in}, PSIG	100	200	250	400	600	850	1250	1450	1600
T_{in}, F	Sat	Sat	550	750	825	900	950	1000	1000
Exhaust, P									
1" HGA	10.20	9.17	8.09	6.85	6.34	5.92	5.62	5.43	5.40
2" HGA	11.31	10.02	8.78	7.36	6.76	6.28	5.94	5.73	5.69
3" HGA	12.12	10.62	9.27	7.71	7.05	6.53	6.16	5.93	5.89
0 PSIG	22.73	17.52	14.57	11.19	9.82	8.81	8.10	7.72	7.62
5 PSIG	26.07	19.35	15.90	11.99	10.42	9.29	8.49	8.07	7.96
10 PSIG	29.52	21.10	17.15	12.71	10.96	9.71	8.83	8.38	8.26
15 PSIG	33.20	22.83	18.35	13.38	11.44	10.08	9.14	8.66	8.52
20 PSIG	37.17	24.56	19.53	14.02	11.90	10.43	9.42	8.91	8.76
25 PSIG	41.56	26.31	20.70	14.63	12.34	10.76	9.68	9.14	8.98
50 PSIG	74.80	35.99	26.75	17.56	14.31	12.22	10.80	10.15	9.94
100 PSIG		66.60	42.40	23.86	18.07	14.77	12.65	11.78	11.46
150 PSIG			71.80	31.93	22.15	17.33	14.35	13.26	12.79
200 PSIG				43.15	26.96	20.05	16.05	14.72	14.08
300 PSIG					40.65	26.53	19.66	17.74	16.70
400 PSIG					78.30	35.43	23.82	21.10	19.52
500 PSIG						49.03	28.87	25.03	22.69
600 PSIG						73.10	35.30	29.79	26.35

Table 5

Theoretical steam rates, lb/KWH

The equation for the theoretical steam rate is as follows:

$$TSR. = 3413/(h_1 - h_2) \qquad (eq\ 1)$$

where:

TSR. = theoretical steam rate of the turbine, lb/kWh

h_1 = throttle enthalpy at the throttle pressure and temperature, Btu/lb

h_2 = extraction or exhaust enthalpy at the exhaust pressure based on isentropic expansion, Btu/1b.

1.8.2 TURBINE GENERATOR ENGINE EFFICIENCY. The engine efficiency is an overall efficiency and includes the entire performance and mechanical and electrical losses of the turbine and generator. The engine efficiency can be calculated using the following equation:

$$n_e = (h_1 - h_e)n_t n_g/(h_1 - h_2) \qquad \text{(eq 2)}$$

where:

n_e = Turbine generator engine efficiency

h_1 and h_2 = (see Equation 1)

h_e = Actual extraction or exhaust enthalpy, Btu/lb

n_t = Turbine mechanical efficiency

n_g = Generator efficiency

Engine efficiency is usually obtained from turbine generator manufacturers or their literature. Therefore, it is not usually necessary to calculate engine efficiency.

1.8.3 ACTUAL STEAM RATE. The actual steam rate of a turbine can be determined by dividing the actual throttle steam flow rate in pounds per hour by the actual corresponding kilowatts, at the generator terminals, produced by that amount of steam. The resulting steam rate is expressed in pounds of steam per kWh. The actual steam rate can also be determined by dividing the theoretical steam rate by the engine efficiency of the turbine generator.

$$ASR = TSR./n_e \qquad \text{(eq 3)}$$

where:

ASR = actual steam rate of the turbine, lb/kWh.

2. COGENERATION IN STEAM POWER PLANTS. Cogeneration in a steam power plant affects the design of the steam turbine relative to the type of cycle used, the exhaust or extraction pressures required, the loading of the steam turbine, and the size of the steam turbine.

2.1 DEFINITION. In steam power plant practice, cogeneration normally describes an arrangement whereby high pressure steam is passed through a turbine prime mover to produce electrical power, and thence from the turbine exhaust (or extraction) opening to a lower pressure steam (or heat) distribution system for general heating, refrigeration, or process use.

2.2 COMMON MEDIUM. Steam power cycles are particularly applicable to cogeneration situations because the actual cycle medium, steam, is also a convenient medium for area distribution of heat.

a) The choice of the steam distribution pressure should be a balance between the costs of distribution, which are slightly lower at high pressure, and the gain in electrical power output by selection of a lower turbine exhaust or extraction pressure.

b) Often, the early selection of a relatively low steam distribution pressure is easily accommodated in the design of distribution and utilization systems, whereas the hasty selection of a relatively high steam distribution pressure may not be recognized as a distinct economic penalty on the steam power plant cycle.

c) Hot water heat distribution may also be applicable as a district heating medium with the hot water being cooled in the utilization equipment and returned to the power plant for reheating in a heat exchange with exhaust (or extraction) steam.

2.3 RELATIVE ECONOMY. When the exhaust (or extraction) steam from a cogeneration plant can be utilized for heating, refrigeration, or process purposes in reasonable phase with the required electric power load, there is a marked economy of fuel energy because the major condensing loss of the conventional steam power plant (Rankine) cycle is avoided. If a good balance can be attained, up to 75 per cent of the total fuel energy can be utilized, as compared with about 40 percent for the best and largest Rankine cycle plants and about 25 to 30 percent for small Rankine cycle systems.

2.4 CYCLE TYPES. The two major steam power cogeneration cycles, which may be combined in the same plant or establishment, are the back pressure and extraction-condensing cycles.

2.4.1 BACK PRESSURE CYCLE. In a back pressure turbine, the entire flow to the turbine is exhausted (or extracted) for heating steam use. This cycle is more effective for heat economy and for relatively lower cost of turbine equipment, because the prime mover is smaller and simpler and requires no condenser and circulating water system. Back pressure turbine generators are limited in electrical output by the amount of exhaust steam required by the heat load and are often governed by the exhaust steam load. They, therefore, usually operate in electrical parallel with other generators.

2.4.2 EXTRACTION-CONDENSING CYCLE. Where the electrical demand does not correspond to the heat demand, or where the electrical load must be carried at times of very low (or zero) heat demand, then condensing-controlled extraction steam turbine prime movers, as shown in Figure 2, may be applicable. Such a turbine is arranged to carry a specified electrical capacity either by a simple condensing cycle or a combination of extraction and condensing. While very flexible, the extraction machine is relatively complicated, requires complete condensing and heat rejection equipment, and must always pass a critical minimum flow of steam to its condenser to cool the low pressure buckets.

2.5 CRITERIA FOR COGENERATION. For minimum economic feasibility, cogeneration cycles will meet the following criteria:

2.5.1 LOAD BALANCE. There should be a reasonably balanced relationship between the peak and normal requirements for electric power and heat. The peak/normal ratio should not exceed 2:1.

2.5.2 LOAD COINCIDENCE. There should be a fairly high coincidence, not less than 70 percent, of time and quantity demands for electrical power and heat.

2.5.3 SIZE. While there is no absolute minimum size of steam power plant which can be built for cogeneration, a conventional steam (cogeneration) plant will be practical and economical only above some minimum size or capacity, below which other types of cogeneration, diesel, or gas turbine become more economical and convenient.

2.5.4 DISTRIBUTION MEDIUM. Any cogeneration plant will be more effective and economical if the heat distribution medium is chosen at the lowest possible steam pressure or lowest possible hot water temperature. The power energy delivered by the turbine is highest when the exhaust steam pressure is lowest. Substantial cycle improvement can be made by selecting an exhaust steam pressure of 40 psig (276 kPa gage) rather than 125 psig (862 kPa gage), for example. Hot water heat distribution should also be considered where practical or convenient, because hot water temperatures of 200 to 240 degrees F (93 to 116 degrees C) can be delivered with exhaust steam pressure as low as 20 to 50 psig (138 to 345 kPa gage). The balance between distribution system and heat exchanger costs, and power cycle effectiveness should be optimized.

3. TURBINE TYPES

3.1 CONDENSING TYPES

3.1.1 HIGH PRESSURE EXTRACTION TYPE. Turbines with throttle pressures generally above 400 psig (2758 kPa gage) are considered high pressure machines; however, the exact demarcation between high, intermediate, and low pressure turbines is not definite. Turbines built with provisions for extraction of steam from the turbine at intermediate pressure points below the throttle pressure are called extraction turbines. The extracted steam may be used for process systems, feed water heating, and environmental heating. A typical cycle using a high pressure extraction type turbine is shown in Figure 2.

3.1.2 HIGH PRESSURE NON-EXTRACTION TYPE. The high pressure non-extraction type of turbine is basically the same as the extraction type described above, except no steam is extracted from the turbine. High pressure steam enters the turbine throttle and expands through the turbine to the condenser. The condenser pressure is comparable to that with high pressure extraction machines.

3.1.3 AUTOMATIC EXTRACTION TYPE. Automatic extraction turbines usually operate with high pressure, high temperature throttle steam supply to a high pressure turbine section. The exhaust pressure of the high pressure turbine is held constant by means of automatic extraction gear (valve) that regulates the amount of steam passing to the low pressure turbine. Single automatic extraction turbines provide steam at a constant pressure from the automatic extraction opening, usually in the range of 50 to 150 psig (345 to 1034 kPa gage). Double automatic extraction turbines consist of a high, intermediate, and low pressure turbine section and provide steam in the range of 50 to 150 psig (345 to 1034 kPa gage) at one automatic extraction opening and 10 to 15 psig (69 to 103 kPa gage) at the other automatic extraction opening. Automatic extraction turbine generators operating automatically meet both automatic extraction steam and electrical demands by adjusting the flow of steam through the low pressure turbine. A typical automatic extraction cycle is shown in Figure 8. Automatic extraction turbines may be either condensing (condenser pressure 1.0 to 4.0 inches of Hg Abs.) or noncondensing (usually 5 to 15 psig (34 to 103 kPa gage) back pressure).

3.1.4 MIXED PRESSURE OR INDUCTION TYPE. The mixed pressure or induction type turbine is supplied with steam to the throttle and also to other stages or sections at a pressure lower than throttle pressure. This type of machine is also called an admission type. The steam admitted into the lower pressure openings may come from old low pressure boilers, or it may be the excess from auxiliary equipment or processes. The mixed pressure turbine is the same as an automatic extraction turbine described above, except steam is admitted instead of extracted at the automatic controlled opening.

3.1.5 LOW PRESSURE TYPE. Low pressure turbines are those with throttle pressures generally below 400 psig (2758 kPa gage). However, the pressure dividing point varies, depending on the manufacturer and type of turbine (industrial, mechanical drive, etc.).

3.2 NONCONDENSING TYPES

3.2.1 SUPERPOSED OR TOPPING TYPE. Refer to Figure 4 in this publication for a description of topping turbine and cycle.

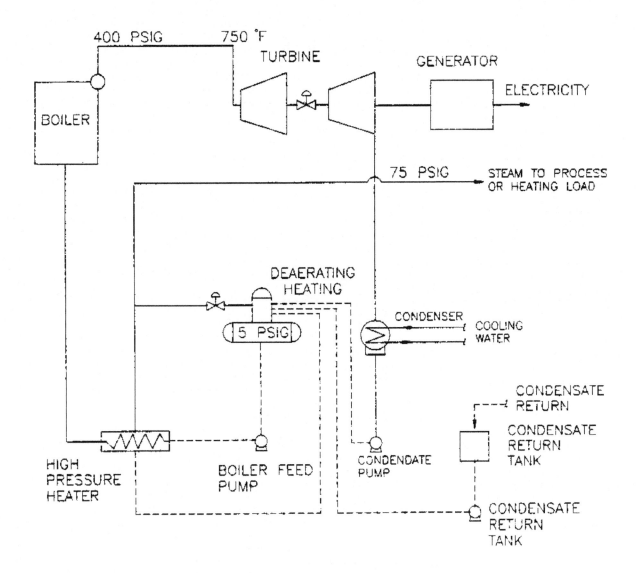

Figure 8
Typical automatic extraction cycle

3.2.2 BACK PRESSURE TYPE. Back pressure turbines usually operate with high pressure, high temperature throttle steam supply, and exhaust at steam pressures in the range of 5 to 300 psig (34 to 2068 kPa gage). Un-controlled steam extraction openings can be provided depending on throttle pressure and exhaust pressures. Two methods of control are possible. One of the methods modulates the turbine steam flow to be such as to maintain the turbine exhaust pressure constant and, in the process, generate as much electricity as possible from the steam passing through the turbine. The amount of electricity generated, therefore, changes upward or downward with like changes in steam demand from the turbine exhaust. A typical back pressure cycle is shown in Figure 3. The other method of control allows the turbine steam flow to be such as to provide whatever power is required from the turbine by driven equipment. The turbine exhaust steam must then be used, at the rate flowing through the turbine, by other steam consuming equipment or excess steam, if any, must be vented to the atmosphere.

3.2.3 ATMOSPHERIC EXHAUST. Atmospheric exhaust is the term applied to mechanical drive turbines which exhaust steam at pressures near atmospheric. These turbines are used in power plants to drive equipment such as pumps and fans.

4. TURBINE GENERATOR SIZES. See Table 6 for nominal size and other characteristic data for turbine generator units.

4.1 NONCONDENSING AND AUTOMATIC EXTRACTION TURBINES. The sizes of turbine generators and types of generator cooling as shown in Table 9 generally apply also to these types of turbines.

4.2 GEARED TURBINE GENERATOR UNITS. Geared turbine generator units utilizing multistage mechanical drive turbines are available in sizes ranging generally from 500 to 10,000 kW. Single stage geared units are available in sizes from 100 kW to 3,000 kW. Multistage units are also available as single valve or multi-valve, which allows further division of size range. Because of overlapping size range, the alternative turbine valve and stage arrangements should be considered and economically evaluated within the limits of their capabilities.

5. TURBINE THROTTLE PRESSURE AND TEMPERATURE. Small, single stage turbines utilize throttle steam at pressures from less than 100 psig (689 kPa gage) and saturated temperatures up to 300 psig and 150 (66 degrees C) to 200 degrees F (93 degrees C) of superheat. Steam pressures and temperatures applicable to larger multistage turbines are shown in Table 7.

5.1 SELECTION OF THROTTLE PRESSURE AND TEMPERATURE. The selection of turbine throttle pressure and temperature is a matter of economic evaluation involving performance of the turbine generator and cost of the unit including boiler, piping, valves, and fittings.

Turbine type and exhaust flow	Nominal last stage blade length, in	Nominal turbine size, kW	Typical generator cooling
Non-reheat units			
Industrial sized			
SCSF	6	2,500	Air
SCSF	6	3,750	Air
SCSF	7	5,000	Air
SCSF	7	6,250	Air
SCSF	8.5	7,500	Air
SCSF	10	10,000	Air
SCSF	11.5	12,500	Air
SCSF	13	15,000	Air
SCSF	14	20,000	Air
SCSF	17-18	25,000	Air
SCSF	20	30,000	Hydrogen
SCSF	23	40,000	Hydrogen
SCF	25-26	50,000	Hydrogen
Utility sized			
TCDF	16.5-18	60,000	Hydrogen
TCDF	20	75,000	Hydrogen
TCDF	23	100,000	Hydrogen
Reheat units (Reheat is never offered for turbine-generators less than 50 MW).			
TCSF	23	60,000	Hydrogen
TCSF	25-26	75,000	Hydrogen
TCDF	16.5-18	100,000	Hydrogen

SCSF = Single case, single flow exhaust
TCSF = Tandem compound single flow exhaust
TCDF = Tandem compound double flow exhaust

Table 6

Direct connected condensing steam turbine generator units

Unit size, kW	Pressure range, psig	Temperature range, deg F
2,500 to 6,250	300 – 400	650 – 825
7,500 to 15,000	500 – 600	750 – 825
20,000 to 30,000	750 – 850	825 – 900
40,000 to 50,000	1,250 – 1,450	825 – 1,000
60,000 to 125,000	1,250 – 1,450	950 – 1,000 and 1,000 reheat

Table 7

Turbine throttle steam pressures and temperatures

5.2 ECONOMIC BREAKPOINTS. Economic breakpoints exist primarily because of pressure classes and temperature limits of piping material that includes valves and fittings. General limits of steam temperature are 750 F (399 degrees C) for carbon steel, 850 degrees F (454 degrees C) for carbon molybdenum steel, 900 degrees F (482 degrees C) for 1/2 to 1 percent chromium - 1/2 percent molybdenum steel, 950 degrees F (510 degrees C) for 1-1/4 percent chromium - 1/2 percent molybdenum steel, and 1,000 degrees F (538 degrees C) for 2-1/4 percent chromium - 1 percent molybdenum. Throttle steam temperature is also dependent on moisture content of steam existing at the final stages of the turbine. Moisture content must be limited to not more than 10 percent to avoid excessive erosion of turbine blades. Traditional throttle steam conditions which have evolved and are in present use are shown in Table 8.

6. TURBINE EXHAUST PRESSURE. Typical turbine exhaust pressure is as shown in Table 9. The exhaust pressure of condensing turbines is dependent on available condenser cooling water inlet temperature.

Pressure, psig	Temperature, degrees F
250	500 or 550
400	650 or 750
600	750 or 825
850	825 or 900
1,250	900 or 950
1,450	950 or 1,000
1600	1,000

Table 8

Typical turbine throttle steam pressure-temperature conditions

Turbine type	Condensing, In Hg Abs	Non-condensing, psig
Multivalve multistage	0.5 – 4.5	0 – 300
Superposed (topping)		200 – 600
Single valve multistage	1.5 – 4.0	0 – 300
Single valve single stage	2.5 – 3.0	1 – 100
Back pressure		5 – 300
Atmospheric pressure		0 - 50

Table 9

Typical turbine exhaust pressure

7. LUBRICATING OIL SYSTEMS

7.1 SINGLE STAGE TURBINES. The lubricating oil system for small, single stage turbines is self-contained, usually consisting of water jacketed, water-cooled, rotating ring-oiled bearings.

7.2 MULTISTAGE TURBINES. Multistage turbines require a separate pressure lubricating oil system consisting of oil reservoir, bearing oil pumps, oil coolers, pressure controls, and accessories.

a) The oil reservoir's capacity shall provide a 5 to 10 minute oil retention time based on the time for a complete circuit of all the oil through the bearings.

b) Bearing oil pump types and arrangement are determined from turbine generator manufacturers' requirements. Turbine generators should be supplied with a main oil pump integral on the turbine shaft. This arrangement is provided with one or more separate auxiliary oil pumps for startup and emergency backup service. At least one of the auxiliary oil pumps shall be separately steam turbine driven or DC motor driven. For some hydrogen cooled generators, the bearing oil and hydrogen seal oil are served from the same pumps.

c) Where separate oil coolers are necessary, two full capacity, water cooled oil coolers shall be used. Turbine generator manufacturers' standard design for oil coolers is usually based on a supply of fresh cooling water at 95 degrees F (35 degrees C) at 125 psig (862 kPa gage). These design conditions shall be modified, if necessary, to accommodate actual cooling water supply conditions. Standard tube material is usually inhibited admiralty or 90-10 copper-nickel. Other tube materials are available, including 70-30 copper-nickel, aluminum-brass, arsenical copper, and stainless steel.

7.3 OIL PURIFIERS. Where a separate turbine oil reservoir and oil coolers are used, a continuous bypass purification system with a minimum flow rate per hour equal to 10 percent of the turbine oil capacity shall be used. Refer to ASME Standard LOS-1M, ASTM-ASME-NEMA Recommended Practices for the Cleaning, Flushing, and Purification of Steam and Gas Turbine Lubricating Systems. The purification system shall be either one of the following types.

7.3.1 CENTRIFUGE WITH BYPASS PARTICLE SIZE FILTER. See Figure 9 for arrangement of equipment. Because of the additives contained in turbine oils, careful selection of the purification equipment is required to avoid the possibility of additive removal by use of certain types of purification equipment such as clay filters or heat and vacuum units. Both centrifuge and particle size filters are suitable for turbine oil purification. Particle filters are generally sized for not less than 5 microns to avoid removal of silicone foam inhibitors if present in the turbine oil used. The centrifuge is used periodically for water removal from the turbine oil. The particle filter, usually of the cellulose cartridge type, is used continuously except during times the centrifuge is used.

7.3.2 MULTISTAGE OIL CONDITIONER. See Figure 10 for arrangement of equipment. The typical multistage conditioner consists of three stages: a precipitation compartment where gross free water is removed by detention time and smaller droplets are coalesced on hydrophobic screens, a gravity filtration compartment containing a number of cloth-covered filter elements, and a storage compartment which contains a polishing filter consisting of multiple cellulose cartridge filter elements. The circulating pump receives oil from the storage compartment and pumps the oil through the polishing filter and back to the turbine oil reservoir. The storage compartment must be sized to contain the flowback oil quantity contained in the turbine generator bearings and oil supply piping. The oil conditioner in this type of purification system operates continuously.

7.4 LUBRICATING OIL STORAGE TANKS. As a minimum, provide one storage tank and one oil transfer pump. The storage tank capacity should be equal to, or greater than the largest turbine oil reservoir. The transfer pump is used to transfer oil between the turbine oil reservoir and the storage tank. The single tank can be used to receive oil from, or return oil to the turbine oil reservoir. Usually a separate portable oil filter press is used for oil purification of used oil held in the storage tank. Two storage tanks can be provided when separate tanks are desired for separate storage of clean and used oil. This latter arrangement can also be satisfied by use of a two compartment single tank. Only one set of storage tanks and associated transfer pump is needed per plant. However, it may be necessary to provide an additional oil transfer pump by each turbine oil reservoir, depending on plant arrangement.

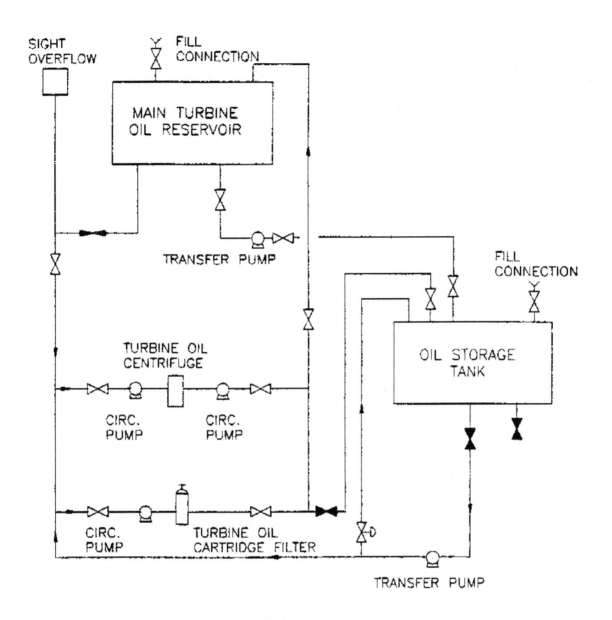

Figure 9

Oil purification system with centrifuge

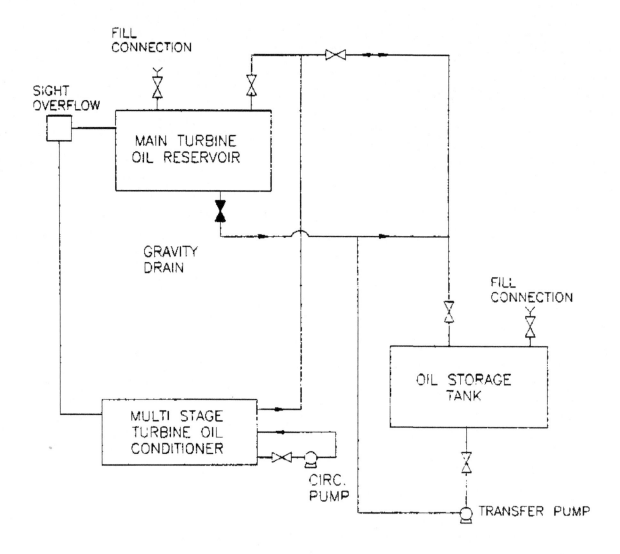

Figure 10
Oil purification system with multistage conditioner

7.5 LUBRICATING OIL SYSTEM CLEANING. Refer to ASME Standard LOS-1M.

8. GENERATOR TYPES. Generators are classified as either synchronous (AC) or direct current (DC) machines. Synchronous generators are available for either 60 cycles (usually used in U.S.A.) or 50 cycles (frequently used abroad). Direct current generators are used for special applications requiring DC current in small quantities and not for electric power production.

9. GENERATOR COOLING

9.1 SELF VENTILATION. Generators, approximately 2,000 kVA and smaller, are air cooled by drawing air through the generator by means of a shaft-mounted propeller fan.

9.2 AIR COOLED. Generators, approximately 2,500 kVA to 25,000 kVA, are air cooled with water cooling of air coolers (water-to-air heat exchangers) located either horizontally or vertically within the generator casing. Coolers of standard design are typically rated for 95 degrees F (35 degrees C) cooling water at a maximum pressure of 125 psig (862 kPa gage) and supplied with 5/8-inch minimum 18 Birmingham wire gage (BWG) inhibited admiralty or 90-10 copper-nickel tubes. Design pressure of 300 psig (2068 kPa gage) can be obtained as an alternate. Also, alternate tube materials such as aluminum-brass, 70-30 copper-nickel, or stainless steel are available.

9.3 HYDROGEN COOLED. Generators, approximately 30,000 kVA and larger, are hydrogen cooled by means of hydrogen to air heat exchangers. The heat exchangers are similar in location and design to those for air-cooled generators. Hydrogen pressure in the generator casing is typically 30 psig (207 kPa gage).

10. TURBINE GENERATOR CONTROL. For turbine generator control description, see the technical literature.

11. TURNING GEAR. In order to thermally stabilize turbine rotors and avoid rotor warpage, the rotors of turbine generators size 12,500 kW and larger are rotated by a motor-driven turning gear at a speed of approximately 5 rpm immediately upon taking the turbine off the line. The rotation of the turbine generator rotor by the turning gear is continued through a period of several hours to several days, depending on the size of the turbine and the initial throttle temperature, until the turbine shaft is stabilized. The turning gear and turbine generator rotor are then stopped until the turbine generator is about to be again placed in service. Before being placed in service, the turbine generator rotor is again stabilized by turning gear rotation for several hours to several days, depending on the turbine size. Turbine generators smaller than 12,500 kW are not normally supplied with a turning gear, since the normal throttle steam temperature is such that a turning gear is not necessary. However, should a turbine be selected for operation at higher than usual throttle steam temperature, a turning gear would be supplied. During turning gear operation, the turbine generator bearings are lubricated by use of either the main bearing oil pump or a separate turning gear oil pump, depending on size and manufacturer of the turbine generator.

12. TURBINE GENERATOR FOUNDATIONS. Turbine generator foundations shall be designed in accordance with the technical literature.

13. AUXILIARY EQUIPMENT. For description of steam jet air ejectors, mechanical air exhausters, and steam operated hogging ejectors, see the technical literature.

14. INSTALLATION. Instructions for turbine generator installation are definitive for each machine and for each manufacturer. For turbine generators, 2,500 kW and larger, these instructions shall be specially prepared for each machine by the turbine generator manufacturer and copies (usually up to 25 copies) shall be issued to the purchaser. The purchase price of a turbine generator shall include technical installation, start-up, and test supervision furnished by the manufacturer at the site of installation.

15. CLEANUP, STARTUP, AND TESTING

15.1 PIPE CLEANING

15.1.1 BOILER CHEMICAL BOIL OUT. Chemical or acid cleaning is the quickest and most satisfactory method for the removal of water side deposits. Competent chemical supervision should be provided, supplemented by consultants on boiler-water and scale problems during the chemical cleaning process. In general, four steps are required in a complete chemical cleaning process for a boiler.

a) The internal heating surfaces are washed with an acid solvent containing a proper inhibitor to dissolve the deposits completely or partially and to disintegrate them.

b) Clean water is used to flush out loose deposits, solvent adhering to the surface, and soluble iron salts. Any corrosive or explosive gases that may have formed in the unit are displaced.

c) The unit is treated to neutralize and "passivate" the heating surfaces. The passivation treatment produces a passive surface or forms a very thin protective film on ferrous surfaces so that formation of "after-rust" on freshly cleaned surfaces is prevented.

d) The unit is flushed with clean water as a final rinse to remove any remaining loose deposits. The two generally accepted methods in chemical cleaning are continuous circulation and soaking.

e) Continuous Circulation. In the circulation method, after filling the unit, the hot solvent is recirculated until cleaning is completed. Samples of the return solvent are tested periodically during the cleaning. Cleaning is considered complete when the acid strength and the iron content of the returned solvent reach equilibrium indicating that no further reaction with the deposits is taking place. The circulation method is particularly suitable for cleaning once-through boilers, superheaters, and economizers with positive liquid flow paths to assure circulation of the solvent through all parts of the unit.

f) Soaking. In cleaning by the soaking method after filling with the hot solvent, the unit is allowed to soak for a period of four to eight hours, depending on deposit conditions. To assure complete removal of deposits, the acid strength of the solvent must be somewhat greater than that required by the actual conditions, since, unlike the circulation method, control testing during the course of the cleaning is not conclusive, because samples of solvent drawn from convenient locations may not truly represent conditions in all parts of the unit. The soaking method is preferable for cleaning units where definite liquid distribution to all circuits by the circulation method is not possible without the use of many chemical inlet connections or where circulation through all circuits at an appreciable rate cannot be assured, except by using a circulating pump of impractical size.

15.1.2 MAIN STEAM BLOWOUT. The main steam lines, reheat steam lines, auxiliary steam lines from cold reheat and auxiliary boiler, and all main turbine seal steam lines shall be blown with steam after erection and chemical cleaning until all visible signs of mill scale, sand, rust, and other foreign substances are blown free. Cover plates and internals for the main steam stop valves, reheat stop, and intercept valves, shall be removed. Blanking fixtures, temporary cover plates, temporary vent and drain piping, and temporary hangers and braces to make the systems safe during the blowing operation shall be installed. After blowing, all temporary blanking fixtures, cover plates, vent and drain piping, valves, hangers, and braces shall be removed. The strainers, valve internals, and cover plates shall be reinstalled. The piping systems, strainers, and valves shall be restored to a state of readiness for plant operation.

15.1.2.1 TEMPORARY PIPING. Temporary piping shall be installed at the inlet to the main turbine and the boiler feed pump turbine to facilitate blowout of the steam to the outdoors. Temporary piping shall be designed in accordance with the requirements of the Power Piping Code, ANSI/ASME B31.1. The temporary piping and valves shall be sized to obtain a cleaning ratio of 1.0 or greater in all permanent piping to be cleaned. The cleanout ratio is determined using the following equation.

$$R = (Q_c/Q_m)^2 \times [(P_v)_c/(P_v)_m \times (P_m/P_c)] \quad \text{(eq 4)}$$

where

R = cleaning ratio
Q_c = flow during cleaning, lb/hr
Q_m = Maximum load flow, lb/hr
$(P_v)_c$ = pressure-specific volume product during cleaning at boiler outlet, ft^3/in^2
$(P_v)_m$ = pressure-specific volume product at maximum load flow at boiler outlet, ft^3/in^2
P_m = pressure at maximum load flow at boiler outlet, psia
P_c = Pressure during cleaning at boiler outlet, psia

This design procedure is applicable to fossil fuel-fired power plants, and is written specifically for drum (controlled circulation) type boilers but may be adapted to once-through (combined circulation) type boilers by making appropriate modifications to the procedure. The same basic concepts for cleaning piping systems apply to all boiler types.

15.1.2.2 BLOWOUT SEQUENCE. Boiler and turbine manufacturers provide a recommended blowout sequence for the main and reheat steam lines. The most satisfactory method for cleaning installed piping is to utilize the following cleaning cycle:

(1) Rapid heating (thermal shock helps remove adhered particles).
(2) High velocity steam blowout to atmosphere.
(3) Thermal cool down prior to next cycle.

The above cycle is repeated until the steam emerging from the blowdown piping is observed to be clean.

15.1.3 INSTALLATION OF TEMPORARY STRAINERS. Temporary strainers shall be installed in the piping system at the suction of the condensate and boiler feed pumps to facilitate removal of debris within the piping systems resulting from the installation procedures. The strainers shall be cleaned during the course of all flushing and chemical cleaning operations. The temporary strainers shall be removed after completion of the flushing and chemical cleaning procedures.

15.1.4 CONDENSER CLEANING. All piping systems with lines to the condenser should be completed and the lines to the condenser flushed with service water. Lines not having spray pipes in the condenser may be flushed into the condenser. Those with spray pipes should be flushed before making the connection to the condenser. Clean the interior of the condenser and hot well by vacuuming and by washing with an alkaline solution and flushing with hot water. Remove all debris. Open the condensate pump suction strainer drain valves and flush the pump suction piping. Prevent flush water from entering the pumps. Clean the pump suction strainers. 5.15.1.5 Condensate System Chemical Cleaning. Systems to be acid and alkaline cleaned are the condensate piping from condensate pump to deaerator discharge, boiler feedwater piping from deaerator to economizer inlet, feedwater heater tube sides, air preheat system piping, and chemical cleaning pump suction and discharge piping. Systems to be alkaline cleaned only, are the feedwater heater shell sides, building heating heat exchanger shell sides, and the feedwater heater drain piping. The chemicals and concentrations for alkaline cleaning are 1000 mg/L disodium phosphate, 2,000 mg/L trisodium phosphate, non-foaming wetting agent as required, and foam inhibitors as required. The chemicals and concentrations for acid cleaning are 2.0 percent hydroxyacetic acid, 1.0 percent formic acid, 0.25 percent ammonium bifluoride, and foaming inhibitors and wetting agents as required.

15.1.4.1 DEAERATOR CLEANING. Prior to installing the trays in the deaerator and as close to unit start-up as is feasible, the interior surfaces of the deaerator and deaerator storage tank shall be thoroughly cleaned to remove all preservative coatings and debris.

Cleaning shall be accomplished by washing with an alkaline service water solution and flushing with hot service water. The final rinse shall be with demineralized water. After cleaning and rinsing, the deaerator and deaerator storage tank shall be protected from corrosion by filling with treated demineralized water.

15.1.4.2 CYCLE MAKEUP AND STORAGE SYSTEM. The cycle makeup and storage system, condensate storage tank, and demineralized water storage tank shall be flushed and rinsed with service water. The water storage tanks should require only a general hose washing. The makeup water system should be flushed until the flush water is clear. After the service water flush, the cycle makeup and storage system shall be flushed with demineralized water until the flushing water has a clarity equal to that from the demineralizer.

15.1.4.3 CONDENSATE-FEEDWATER AND AIR PREHEAT SYSTEMS. The condensate-feedwater and air preheat systems (if any) shall be flushed with service water. The condensate pumps shall be used for the service water flushing operations. Normal water level in the condenser should be maintained during the service water flushing operation by making up through the temporary service water fill line. After the service water flush, the condensate-feedwater and air preheat systems shall be flushed with demineralized water. After the demineralized water flush, the condensate-feedwater and air preheat systems shall be drained and refilled with demineralized water.

15.1.4.4 ALKALINE CLEANING. The condensate-feedwater and air preheat water systems shall be alkaline cleaned by injecting the alkaline solution into circulating treated water, preheated to 200 degrees F (93.3 degrees C) by steam injection, until the desired concentrations are established. The alkaline solution should be circulated for a minimum of 24 hours with samples taken during the circulation period. The samples should be analyzed for phosphate concentration and evidence of free oil. The feedwater heaters and drain piping shall be alkaline cleaned by soaking with hot alkaline cleaning solution in conjunction with the condensate-feedwater and air preheat water system alkaline cleaning. The heater shells and drain piping should be drained once every six hours during the circulation of the alkaline cleaning solution through the condensate feedwater and air preheat water systems. After the alkaline cleaning is completed, flush the condensate-feedwater, air preheat water, feedwater heater, and drain piping systems with demineralized water.

15.1.4.5 ACID CLEANING. Acid cleaning of the condensate feedwater and air preheat water systems shall be similar to the alkaline cleaning, except that the circulation period shall only be six hours. The condensate-feedwater and air preheat water system shall be heated to 200 degrees F (93 degrees C) and hydrazine and ammonia injected into the circulating water to neutralize the acid solution. The systems shall then be flushed with demineralized water until all traces of acid are removed.

15.1.6 TURBINE LUBE OIL FLUSH AND RECIRCULATION. The lubricating and seal oil systems of the turbine generator shall be cleaned as recommended by the manufacturer. Oil samples shall be tested to determine contamination levels. The cleaning shall be a cold flushing of the system and cleaning of the oil reservoir. This shall be followed by cycling of circulating hot and cold oil until the system is clean.

15.2 EQUIPMENT STARTUP

15.2.1 PRELIMINARY CHECKS. Preliminary checks and inspection, and any required corrective work shall be performed on all equipment in accordance with the equipment manufacturer's recommendations.

15.2.1.1 SHAFT ALIGNMENT. All bearings, shafts, and other moving parts shall be checked for proper alignment.

15.2.1.2 LINKAGE ALIGNMENT. Manual set of all linkages shall be performed, ensuring open and close limit adjustment. Operational linkage adjustment shall be performed as required.

15.2.1.3 SAFETY EQUIPMENT. All coupling guards, belt guards, and other personnel safety items shall be installed.

15.2.1.4 PIPING. All power actuated valves shall be checked for correct valve action and seating and the actuators and converters shall be given initial adjustment. All manual valves shall be operated to ensure correct operation and seating. All safety valves shall be checked for correct settings. All piping shall be nondestructively tested, hydrostatically tested, leak tested, or air tested, as applicable, and shall be flushed or blown clean. All temporary shipping braces, blocks, or tie rods shall be removed from expansion joints. All spring type pipe hangers shall be checked for proper cold settings.

15.2.1.5 PITS. All pump suction pits shall be free of trash.

15.2.1.6 LUBRICATION. Each lubricating oil system shall be flushed and the filters inspected. All oil tanks, reservoirs, gear cases, and constant level type oilers shall be checked for proper oil levels. All points requiring manual lubrication shall be greased or oiled as required.

15.2.1.7 BELTS, PULLEYS, AND SHEAVES. All belts, pulleys, and sheaves shall be checked for correct alignment and belt tension.

15.2.1.8 COOLING AND SEALING WATER. All cooling and sealing water supplies shall be flushed and checked for proper operation.

15.2.1.9 PUMP SUCTION STRAINERS. All pump suction strainers shall be installed.

15.2.1.10 STUFFING BOXES AND PACKING. All stuffing boxes shall be checked for correct takeup on the packing.

15.2.1.11 MECHANICAL SEALS. All mechanical seals shall be removed as required to ensure clean sealing surfaces prior to starting. Seal water piping shall be cleaned to the extent necessary to ensure no face contamination. Seal adjustments shall be performed as required by the manufacturer.

15.2.1.12 TANKS AND VESSELS. All tanks and vessels shall be thoroughly inspected internally before securing.

15.2.2 INITIAL PLANT STARTUP. The following steps shall be followed for plant startup:

- Operate demineralizer and fill condensate storage/return tank.
- Fill boiler, deaerator, and condenser.
- Start boiler feed pumps.
- Warm up boiler using manufacturer's recommendations.
- Start cooling water system pumps.
- Start condensate pumps.
- Start condenser exhauster (air ejectors).
- Start turbine lubricating oil system.
- Roll turbine using manufacturer's startup procedures.

15.3 TESTING. For testing requirements, see the technical literature.

16. OPERATION

16.1 TRIAL OPERATION. After all preliminary checks and inspections are completed, each piece of equipment shall be given a trial operation. Trial operation of all equipment and systems shall extend over such period of time as is required to reveal any equipment weaknesses in bearings, cooling systems, heat exchangers, and other such components, or any performance deficiencies which may later handicap the operation of main systems and the complete plant. All rotating equipment shall be checked for overheating, noise, vibration, and any other conditions which would tend to shorten the life of the equipment.

16.2 MAIN SYSTEM OPERATION. Main systems should be trial operated and tested after each individual piece of equipment has been trial operated and ready for operation. All functional and operational testing of protective interlocking, automatic controls, instrumentation, alarm systems, and all other field testing should be conducted during initial plant startup. All piping should be visually inspected for leaks, improper support adjustment, interferences, excessive vibration, and other abnormal conditions. Steam traps should be verified for proper operation and integral strainers cleaned.

16.3 OPERATION CONTROL. A system of control to protect personnel and equipment as the permanent plant equipment and systems are completed and capable of energization, pressurization, or being operated, should be established. The system should consist of placing appropriate tags on all equipment and system components. Tags should indicate status and the mandatory clearances required from designated personnel to operate, pressurize, energize, or remove from service such equipment or systems. The controls established should encompass the following phases.

16.3.1 EQUIPMENT OR SYSTEMS COMPLETED to the point where they may be energized, pressurized, or operated, but not yet checked out, shall be tagged. The sources of power or pressure shall be turned off and tagged.

16.3.2 EQUIPMENT AND SYSTEMS RELEASED for preoperational check-out shall be so tagged. When a request to remove from service is made, all controls and sources of power or pressure shall be tagged out and shall not be operated under any circumstances.

Printed in Great Britain
by Amazon